Mastering your iphone 15 pro max guide for seniors

A Comprehensive User manual for Unlock the Full Potential of Your apple 15 pro max with step by step instructions of Expert Tips and Tricks.

Carlos D.Gray

Copyright page

Tables of contents

Chapter 1: Introduction

Welcome to "Mastering Your iPhone 15 Pro Max: A Comprehensive User Guide." This book is your go-to resource for navigating the advanced features of the iPhone 15 Pro Max and making the most of its capabilities.

In this guide, we'll walk you through the essentials, from setting up your device to mastering the intricacies of its powerful features. Whether you're a tech enthusiast or a casual user, this book is designed to be your companion in unraveling the full potential of your iPhone.

Discover how to personalize your experience, explore advanced camera functionalities, troubleshoot common issues, and stay connected with the latest in Apple's innovative technology. Let's embark on a journey to empower you with the knowledge and skills needed to truly unlock the possibilities within your iPhone 15 Pro Max.

Welcome once again to the digital realm where innovation meets sophistication – welcome to "Mastering Your iPhone 15 Pro Max: A Comprehensive User Guide." In a world driven by technology, your iPhone 15 Pro Max stands as a testament to cutting-edge design and functionality. This chapter serves as your gateway to understanding the essence of this guide, providing you with an overview of what awaits within these pages.

Overview of iPhone 15 Pro Max

At the heart of the iPhone 15 Pro Max lies a fusion of art and technology. The device boasts a stunning Super Retina XDR display, offering a visual feast with vibrant colors and sharp contrasts. The A15 Bionic chip, a powerhouse beneath the sleek exterior, ensures seamless performance, whether you're navigating through apps, indulging in multimedia, or diving into augmented reality experiences.

The triple-camera system, featuring advanced sensors and computational photography capabilities, transforms every moment into a masterpiece. From capturing intricate details to embracing low-light scenarios, the camera on the iPhone 15 Pro Max is your passport to a world of creative expression.

As we delve deeper, you'll uncover the nuances of the design – the ergonomics that marry form and function. This section is your window into the technological marvel that rests in your hands, emphasizing the features that set the iPhone 15 Pro Max apart and make it a worthy companion in your daily endeavors.

Setting Up Your Device

Now that you have a glimpse of the technological prowess within your iPhone 15 Pro Max, it's time to embark on the initial journey of making it truly yours. Whether you're unboxing a brand-new device or upgrading from a previous model, the setup process is the gateway to a personalized user experience.

This section takes you by the hand, guiding you through the activation process, language and region settings, and connecting to Wi-Fi. Learn how to seamlessly transfer data from your previous device, ensuring a smooth transition into the enhanced world of the iPhone 15 Pro Max.

Dive into the essentials of Apple ID creation and configuration, a key step in unlocking the full potential of the iOS ecosystem. Understand the importance of enabling features like Face ID or

Touch ID, fortifying your device with an added layer of security while ensuring effortless accessibility.

As you progress through this section, you'll find tips and tricks to streamline the setup process, making it not just a technical necessity but an enjoyable initiation into the world of your iPhone. Get ready to witness your device come to life with each configured setting, setting the stage for an immersive and tailored user experience.

Embarking on this introductory chapter, you're not merely opening a user guide – you're stepping into a realm where technology aligns with your lifestyle. The iPhone 15 Pro Max is more than a device; it's a companion, an enabler, and a canvas for your digital adventures. As we journey through the subsequent chapters, the depth and breadth of your iPhone experience will unfold, empowering you to navigate with confidence and master the nuances of this technological masterpiece.

Chapter 2: Navigating Your iPhone

Now that you've unboxed and set up your iPhone 15 Pro Max, it's time to embark on a journey through the intricacies of navigation, unraveling the layers of its intuitive interface to make your interactions with the device seamless and efficient.

Home Screen Basics

The Home screen is the gateway to your digital world, and understanding its nuances enhances your overall user experience. Learn the art of arranging and organizing apps, creating folders, and customizing your Home screen layout to reflect your priorities and preferences. Discover the simplicity of the App Library, a feature designed to intelligently organize and categorize your apps for easy access.

This section goes beyond the basics, introducing you to the dynamic world of widgets. Widgets provide at-a-glance information right on your Home screen, offering a personalized and interactive experience. Explore the myriad of widgets available, from weather updates to calendar events, and learn how to tailor your Home screen to suit your lifestyle.

Control Center and Notifications

Efficiency meets convenience with the Control Center – a versatile panel that puts essential controls at your fingertips. Uncover its functionalities, from toggling Wi-Fi and Bluetooth to adjusting screen brightness. Dive into customization options, ensuring that the Control Center aligns with your preferences, providing a personalized shortcut to the features you use most frequently.

Notifications play a pivotal role in keeping you informed, but managing them effectively is key to a harmonious digital experience. Delve into the settings to fine-tune app-specific notification preferences, ensuring that you stay connected without being inundated. Learn to leverage "Do Not Disturb" and other features to maintain focus when needed.

Using Face ID and Touch ID

Security and accessibility converge with Face ID and Touch ID, offering seamless and secure methods to unlock your iPhone. Explore the sophisticated facial recognition capabilities of Face ID or the tactile precision of Touch ID, tailoring these biometric authentication methods to your preferences. Understand the settings that enhance the balance between security and convenience, ensuring your device is not only a fortress of protection but also easily accessible in your day-to-day activities.

As you navigate through the layers of your iPhone's interface, you're not just interacting with a device – you're sculpting a personalized digital space. The Home screen becomes a canvas where your favorite apps and widgets converge, the Control Center transforms into a personalized control panel, and biometric authentication adds a layer of security seamlessly integrated into your daily routine.

These foundational skills set the stage for a user-friendly journey, where your iPhone adapts to your preferences, making every interaction a pleasure. Join us in the upcoming chapters as we explore the essential apps and features, building upon the foundation laid in this exploration of your iPhone's navigation.

Chapter 3: Essential Apps and Features

Now that you've mastered the art of navigating your iPhone 15 Pro Max, it's time to delve into the core of its functionality – the essential apps and features that define the iPhone experience. In this chapter, we'll explore the intricacies of the camera, communication through messages, web browsing with Safari, and the process of downloading and managing apps from the App Store.

Camera and Photography Tips

The iPhone 15 Pro Max's camera system is a marvel, equipped with advanced sensors and computational photography capabilities. Let's unlock the full potential of your device's camera and delve into capturing moments with finesse.

Explore the Camera app, understanding its various modes and settings. From Portrait mode for stunning depth-of-field effects to Night mode for exceptional low-light photography, the camera offers a diverse range of tools at your disposal. Learn to optimize your shots with features like Smart HDR, Deep Fusion, and ProRAW, ensuring that every photo is a masterpiece.

Dive into the world of video recording, understanding the capabilities of 4K Dolby Vision HDR recording and cinematic video stabilization. Unleash your creativity with features like Time-lapse and Slo-mo, adding a dynamic touch to your visual storytelling.

This section goes beyond the basics, offering photography tips and tricks. From composition techniques to leveraging natural light, discover how to elevate your photography skills with your iPhone 15 Pro Max. Whether you're a photography enthusiast or a casual shooter, this exploration of the camera features ensures that you capture moments in their full glory.

Messages and Communication

Communication lies at the heart of your iPhone experience, and the Messages app is the conduit through which you connect with friends and family. Let's unravel the features that make messaging on your iPhone a seamless and expressive experience.

Discover the versatility of iMessage, from sending text messages to sharing photos, videos, and even your location. Embrace the expressive power of Memoji and Animoji, adding a personal touch to your messages. Understand how to use reactions, tapbacks, and the inline reply feature to engage in lively conversations.

This section also delves into the world of group messaging, offering tips on managing group chats effectively. From customizing group names to muting notifications, learn to navigate group conversations with ease.

Safari: Browsing Made Easy

Safari is your gateway to the internet, and mastering its features enhances your browsing experience. Explore the streamlined interface, learn to open multiple tabs, and discover how to save your favorite websites as bookmarks for quick access.

Navigate through the settings, understanding how to customize your browsing preferences. From enabling Reader View for distraction-free reading to utilizing content blockers for an ad-free experience, Safari empowers you to tailor your browsing environment.

Discover the power of tab management, exploring the tab switcher, and leveraging features like Tab Groups to organize your browsing sessions efficiently. Whether you're researching, shopping, or simply exploring the web, Safari is your versatile companion, adapting to your browsing habits.

App Store and Downloading Apps

The App Store is a treasure trove of applications that cater to every aspect of your interests and needs. Learn to navigate the App Store, discover trending apps, and explore curated collections that align with your preferences.

Understand the process of downloading and updating apps, ensuring that your device is equipped with the latest features and security enhancements. Delve into the App Library, a feature designed to declutter your Home screen and intelligently organize your apps for easy access.

Explore app management, from rearranging apps to deleting and offloading them to free up storage space. This section also provides insights into app permissions and privacy settings, empowering you to control how apps access your personal information.

As you journey through the essential apps and features, you're not just using your iPhone — you're optimizing your digital lifestyle. The camera becomes a tool for creative expression, Messages transforms into a dynamic communication hub, Safari adapts to your browsing preferences, and the App Store becomes a gateway to a vast digital ecosystem.

These essential skills lay the foundation for a comprehensive iPhone experience, setting the stage for further exploration into customization, advanced features, and troubleshooting in the upcoming chapters. Join us as we continue to unravel the layers of your iPhone 15 Pro Max, empowering you to navigate its capabilities with confidence and flair.

Chapter 4: Customizing Your Experience

Now that you've explored the essential apps and features of your iPhone 15 Pro Max, it's time to take the next step in personalizing your device to align seamlessly with your preferences and lifestyle. In this chapter, we'll delve into customizing settings, creating widgets, and managing storage to tailor your iPhone experience.

Personalizing Settings

The Settings app is the control center for fine-tuning your device to match your unique preferences. Navigate through the various categories, from General to Display, and understand the plethora of customization options available.

Dive into Accessibility settings, unlocking features that cater to different needs and preferences. From adjusting text size and display settings to exploring voice control and sound accommodations, personalize your device to enhance usability.

Explore the intricacies of Sounds & Haptics, shaping the auditory and tactile feedback of your iPhone. Customize ringtones, vibrations, and notification sounds to create a personalized and distinctive experience.

This section also covers Privacy settings, empowering you to control how your device handles personal information. From app permissions to location services, understand how to manage your privacy settings effectively.

Creating Widgets

Widgets are a dynamic way to bring essential information to your Home screen, providing at-a-glance updates without needing to open apps. Explore the Widget Gallery, discovering a variety of widgets that cater to different needs and interests.

Learn how to add, rearrange, and customize widgets on your Home screen, ensuring that the information most relevant to you is readily accessible. Whether it's weather updates, calendar events, or fitness tracking, widgets transform your Home screen into a personalized dashboard.

Delve into the Smart Stack, a feature that intelligently rotates through relevant widgets based on your usage patterns. This dynamic stack ensures that the right information is presented at the right time, enhancing the overall efficiency of your device.

Managing Storage

Efficient storage management is key to maintaining optimal performance on your iPhone. Navigate through the Storage settings, gaining insights into how your device utilizes its storage capacity.

Explore recommendations for optimizing storage, from reviewing large attachments and unused apps to managing media downloads. Understand how to offload unused apps, a feature that preserves app data while freeing up storage space on your device.

This section also provides insights into iCloud storage management, ensuring that your essential data is backed up securely in the cloud. Learn how to set up and manage iCloud backups, ensuring that your important files, photos, and documents are safeguarded.

As you personalize your settings, create widgets, and manage storage, your iPhone transforms from a device into a tailored digital companion. Each adjustment reflects your unique preferences and streamlines your interaction with the device, making it an extension of your personality and lifestyle.

These customization skills lay the groundwork for a truly personalized iPhone experience, setting the stage for further exploration into advanced features, security settings, and troubleshooting in the chapters to come. Join us as we continue to unveil the layers of your iPhone 15 Pro Max, empowering you to navigate its capabilities with confidence and flair.

Chapter 5: Advanced Features

Having laid the foundation for personalization and exploration of essential functionalities, it's time to delve into the advanced features that set your iPhone 15 Pro Max apart. In this chapter, we'll explore the sophisticated capabilities of the Pro Max camera, dive into the realm of augmented reality (AR) applications, and ensure the security and privacy of your device through advanced settings.

Pro Max Camera Features

The Pro Max variant of the iPhone 15 is synonymous with an advanced camera system, elevating your photography and videography experience to new heights. Let's unravel the intricacies of the Pro Max camera features and discover how to capture moments with professional finesse.

Explore the ProRAW capabilities, providing you with unprecedented control over your photos. Understand how to leverage ProRes video recording, ensuring cinematic quality in your video projects. Dive into advanced camera settings, adjusting exposure, focus, and white balance to achieve the perfect shot in any lighting condition.

This section also covers the Pro Max's telephoto lens and LiDAR scanner, unlocking capabilities such as Night mode portraits and enhanced autofocus in low-light scenarios. Learn how to make the most of these advanced features to capture stunning images that stand out.

Augmented Reality (AR) Applications

Augmented Reality transforms your iPhone into a portal where the digital and physical worlds seamlessly blend. Discover the world of AR applications, from gaming experiences to practical tools that enhance your everyday life.

Explore ARKit, Apple's platform for AR development, and understand how developers leverage this technology to create immersive applications. Dive into AR gaming, where your surroundings become the playground for interactive experiences. Discover educational AR apps that bring learning to life, allowing you to explore complex concepts in a tangible and engaging way.

This section also covers practical AR applications, such as measuring objects with the Measure app or visualizing furniture in your living space with IKEA Place. Understand how AR enhances navigation with apps like Apple Maps, providing intuitive overlays of directions directly in your real-world environment.

Security and Privacy Settings

Ensuring the security and privacy of your device is paramount. In this section, we'll navigate through advanced security settings, empowering you to fortify your iPhone 15 Pro Max against potential threats.

Explore the intricacies of Face ID and Touch ID settings, adjusting sensitivity and control features for biometric authentication. Dive into the Face ID attention settings, ensuring that your device unlocks only when you want it to.

Understand the nuances of passcode settings, exploring options for alphanumeric passcodes and automatic wipe settings after failed attempts. Learn how to enable two-factor authentication for an additional layer of security, safeguarding your Apple ID and connected services.

This section also delves into privacy settings, providing insights into app tracking transparency, location services, and managing app permissions. Understand how to review and control which apps have access to sensitive data, ensuring that your privacy remains in your hands.

As you explore the advanced features of your iPhone 15 Pro Max, you're not just using a device – you're immersing yourself in a world of technological sophistication. The Pro Max camera becomes a tool for professional-level creativity, AR applications transform your surroundings into interactive experiences, and security settings fortify your device against potential threats.

These advanced skills pave the way for a comprehensive understanding of your iPhone's capabilities, setting the stage for further exploration into connectivity, iCloud services, and troubleshooting in the upcoming chapters. Join us as we continue to uncover the layers of your iPhone 15 Pro Max, empowering you to navigate its advanced features   with   confidence   and   mastery.

Chapter 6: Troubleshooting and Tips

In the dynamic landscape of technology, occasional challenges may arise, but fear not—this chapter is your guide to troubleshooting common issues and discovering invaluable tips to optimize your iPhone 15 Pro Max experience. Let's explore the intricacies of identifying and resolving challenges while uncovering tips that enhance the efficiency and longevity of your device.

Common Issues and Solutions

Technology, while advanced, can occasionally present challenges. In this section, we'll address common issues users may encounter and provide effective solutions.

One common concern is battery life. Understand how to identify battery-draining apps, optimize settings to conserve energy, and maximize the lifespan of your device's battery. Dive into connectivity issues, learning how to troubleshoot Wi-Fi and Bluetooth problems, ensuring a seamless online and device-to-device experience.

Explore solutions for app-related issues, from crashes to sluggish performance. Learn the art of troubleshooting software glitches through methods like force restarting your iPhone or reinstalling problematic apps. This section ensures that you have the tools to tackle common challenges swiftly.

Maximizing Battery Life

Battery life is a critical aspect of your iPhone experience, and understanding how to maximize it ensures prolonged usage without constant recharging.

Explore settings that impact battery consumption, from adjusting screen brightness to managing background app refresh. Uncover the power of Low Power Mode, a feature designed to conserve energy during critical moments. Learn about battery health, how to check it, and tips for maintaining optimal battery performance over time.

This section goes beyond the basics, providing insights into the impact of location services and push notifications on battery life. Understand how to strike the right balance between functionality and energy efficiency, ensuring that your iPhone is ready whenever you need it.

Software Updates and Maintenance

Keeping your iPhone's software up to date is crucial for optimal performance, security, and access to the latest features. Explore the process of checking for and installing software updates, ensuring that your device benefits from the latest advancements.

Learn about the importance of regular maintenance, from cleaning your device physically to managing storage and decluttering unnecessary files. Understand how to optimize settings, ensuring that your iPhone operates at peak efficiency.

This section also covers the significance of backing up your device regularly. Discover the ease of iCloud backups and learn how to transfer data seamlessly when upgrading to a new iPhone or restoring your current device.

As you navigate through troubleshooting common issues, maximizing battery life, and maintaining your device through software updates, you're not just resolving challenges—you're actively contributing to the longevity and seamless operation of your iPhone 15 Pro Max.

These troubleshooting and maintenance skills set the stage for a resilient and optimized iPhone experience, empowering you to tackle challenges with confidence and ensuring that your device remains a reliable companion in your daily endeavors.

In the upcoming chapters, we'll explore connectivity options, delve into iCloud services, and guide you through seeking help when needed. Join us as we continue to unravel the layers of your iPhone 15 Pro Max, empowering you with the knowledge to navigate and maintain your device with expertise.

Chapter 7: Connectivity and iCloud

Connectivity lies at the core of your iPhone 15 Pro Max experience, facilitating seamless communication, data sharing, and synchronization. In this chapter, we will explore the intricacies of connectivity, from managing Wi-Fi and cellular settings to harnessing the power of iCloud for secure data storage and synchronization.

Wi-Fi and Cellular Settings

Efficiently managing your Wi-Fi and cellular settings ensures a smooth online experience and optimal use of your device's capabilities. In this section, we'll delve into the nuances of connectivity settings, offering insights into Wi-Fi networks, cellular data usage, and ways to optimize your connection.

Explore the Wi-Fi settings, learning how to connect to networks, prioritize preferred networks, and troubleshoot common Wi-Fi issues. Understand the significance of Personal Hotspot, a feature that turns your iPhone into a mobile Wi-Fi hotspot for sharing your cellular data connection with other devices.

Dive into cellular settings, gaining insights into data usage tracking, managing cellular data for specific apps, and optimizing settings to balance performance and efficiency. Learn about the potential of 5G connectivity, ensuring that you harness the full power of high-speed data when available.

This section also covers Airplane Mode and other connectivity toggles, providing you with tools to manage connectivity in various scenarios. Whether you're conserving battery life, avoiding interruptions, or troubleshooting connectivity issues, understanding these settings ensures that your iPhone adapts to your needs.

iCloud Storage and Syncing

ICloud is a cornerstone of the Apple ecosystem, providing a secure and convenient platform for storing and synchronizing your data across devices. In this section, we'll unravel the capabilities of iCloud, from managing storage to leveraging synchronization features.

Explore iCloud storage options, understanding how to upgrade your plan and manage your available space. Delve into iCloud Backup, a crucial feature for safeguarding your essential data. Learn how to initiate backups manually and ensure that your device is always prepared for unexpected situations.

Understand the power of iCloud Photos, a feature that seamlessly syncs your photos and videos across devices, ensuring that your memories are accessible wherever you go. Explore iCloud Drive, a file storage service that enables seamless access to documents, presentations, and other files on multiple devices.

This section also covers iCloud Keychain, a secure password manager that simplifies your online experience by securely storing and autofilling passwords across your Apple devices. Uncover the magic of Find My, a feature that not only helps you locate your device but also enables you to track friends and family who share their location with you.

As you navigate through Wi-Fi and cellular settings, mastering the intricacies of iCloud storage and syncing, you're not just managing connections and data—you're building a seamless digital experience that transcends individual devices.

These connectivity and iCloud skills set the stage for a cohesive Apple ecosystem, ensuring that your iPhone is not an isolated device but a hub that synchronizes seamlessly with your Mac, iPad, and other Apple devices. Join us as we continue to unravel the layers of your iPhone 15 Pro Max, empowering you to navigate its connectivity options and iCloud services with confidence and synchronization mastery.

Chapter 8: Getting Help

In the ever-evolving world of technology, seeking help when needed is a crucial aspect of navigating your iPhone 15 Pro Max effectively. This chapter serves as your guide to the array of Apple support resources, community forums, and troubleshooting assistance that ensure you have the support needed to make the most of your device.

Apple Support Resources

Apple offers a rich tapestry of support resources designed to address a wide range of issues and queries. Whether you're facing technical challenges, exploring advanced features, or seeking guidance on device maintenance, Apple's support resources are there to assist you.

Explore the Apple Support website, a comprehensive hub that hosts guides, articles, and step-by-step instructions for troubleshooting common issues. Discover the power of the Apple Support app, an invaluable tool that provides personalized assistance and allows you to schedule appointments at Apple Stores or authorized service providers.

Dive into the world of AppleCare, Apple's extended warranty and support service. Understand the benefits of AppleCare, from priority access to technical experts to coverage for accidental damage. Learn how to check your device's warranty status and explore available support options tailored to your needs.

Explore the Apple Support Communities, an online forum where users share experiences, insights, and solutions. Engage with the community, seeking advice and contributing your knowledge to create a collaborative space for troubleshooting and learning.

Community Forums and Assistance

The Apple Support Communities provide a unique platform for users to connect, share, and assist each other. Dive into these forums, where a wealth of collective knowledge is at your fingertips.

Understand the etiquette of the forums, ensuring a positive and respectful engagement with the community. Learn how to effectively search for existing threads to find solutions to common issues or start a new discussion to seek personalized assistance.

Explore other online communities, platforms, and social media groups where Apple users gather to discuss their experiences. From Reddit to specialized forums, discover additional resources that cater to specific interests and concerns.

Conclusion: Navigating Your iPhone with Confidence

As we conclude this guide, it's essential to recognize that your iPhone 15 Pro Max is not merely a device but a gateway to a world of possibilities. From mastering the basics of navigation to exploring advanced features, personalizing settings, and troubleshooting challenges, you've embarked on a journey to unlock the full potential of your device.

By understanding the vast support ecosystem provided by Apple, you've equipped yourself with the tools to seek assistance and engage with a community of users who share their knowledge and experiences. Whether you prefer exploring official support channels, participating in forums, or leveraging social media communities, the resources are at your disposal.

Remember, technology evolves, and your iPhone is designed to adapt and grow with you. Regularly explore software updates, stay informed about new features, and continue to refine your understanding of your device's capabilities. As you navigate through the intricacies of your iPhone, you're not just using technology – you're embracing a digital lifestyle that empowers you to connect, create, and explore.

In the ever-evolving landscape of technology, the journey is ongoing, and the possibilities are limitless. Continue to navigate your iPhone with confidence, curiosity, and a sense of discovery. As you do, your iPhone transforms from a device into a personalized digital companion that enhances every aspect of your daily   life.

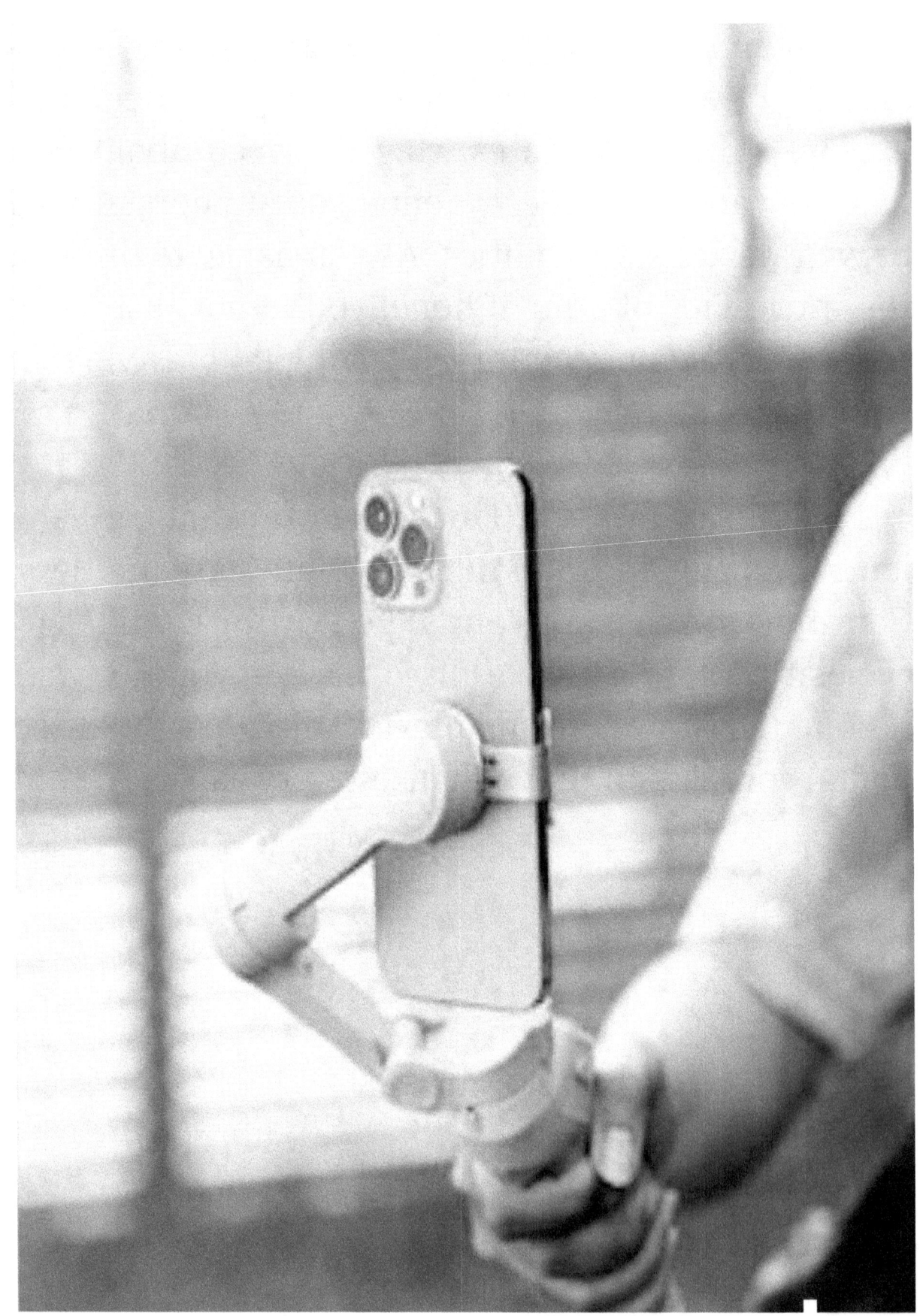